WITHDRAWN

Splitting the Atom

Alan Morton

WORLD ALMANAC® LIBRARY

Please visit our web site at: www.worldalmanaclibrary.com
For a free color catalog describing World Almanac® Library's
list of high-quality books and multimedia programs,
call 1-800-848-2928 (USA) or 1-800-387-3178 (Canada).
World Almanac® Library's Fax: (414) 332-3567.

Library of Congress Cataloging-in-Publication Data

Morton, Alan Q.
       Splitting the atom/ by Alan Morton.
              p. cm. — (Milestones in modern science)
       Includes bibliographical references and index.
       ISBN 0-8368-5856-5 (lib. bdg.)
       1. Atoms. 2. Nuclear physics. 3. Nuclear fission. 4. Particles (Nuclear physics) I. Title.
    II. Series.
    QC776.M67   2005
    539.7—dc22   2005043059

This North American edition first published in 2006 by
World Almanac® Library
A Member of the WRC Media Family of Companies
330 West Olive Street, Suite 100
Milwaukee, WI 53212 USA

This edition © 2006 by World Almanac® Library. First published by Evans Brothers Limited. © 2005 by Evans Brothers Limited.
2A Portman Mansions, Chiltern Street, London W1U 6NR, United Kingdom. This U.S. edition published under license from
Evans Brothers Limited.

Evans Brothers Consultant: Dr. Anne Whitehead
Evans Brothers Editor: Sonya Newland
Evans Brothers Designer: D. R. Ink, info@d-r-ink.com
Evans Brothers Picture researcher: Julia Bird

World Almanac® Library editor: Carol Ryback
World Almanac® Library cover design and art direction: Tammy West

Photo credits: (t) top, (b) bottom, (r) right, (l) left
Science photo library: cover, 12(t), 12(b), 14, 19(t), 21, 27(t); /Los Alamos National Laboratory cover, 27(b), 33 (background), 36;/
CERN cover, 38; /EFDA-JET 3, 35(b); /U.S. Department of Energy 4(t); /Laguna Design 4(b), 44; /Fermilab 5, 40; /Sheila Terry 9(t),
15(t); /Martyn R. Chillmaid 10; /Jean-Loup Charmet 15(b); /Martin Dohrn 16(b); /Mehau Kulyik 23(t); /Dept. of Physics, Imperial
College 11(b); /Lande Collection/American Institute of Physics 23(b); /C. Powell, P. Fowler, and D. Perkins 24(t); /David Parker 25(b), 39;
/Lawrence Berkeley Laboratory 26, 29(b); /Argonne National Laboratory 30(t), 30(b); /Martin Bond 37(t); /Photo © Estate of Francis Bello
42; /ArSciMed 43; /U.S. Air Force 33 (inset); /NASA 34. Big Blue Ltd.: Robert Walster 7(tl), 7(bl). Science & Society Picture Library:
20; /Science Museum 11(t), 22, 29(t); /National Museum of Photography, Film & Television 17; /CERN 41. © CORBIS: 28, 31, 32(b).
CORBIS: /© Archivo Iconografico, S.A. 6; /© Royalty-Free 7(br); /© Stefano Blanchetti 8(t); /© Bettmann 13, 37(b).

Printed in the United States of America

1 2 3 4 5 6 7 8 9 09 08 07 06 05

CONTENTS

Introduction

Above: Atomic bombs work by releasing energy in the form of heat and radiation from the nucleus of atoms. Such weapons are the result of a century of investigating and harnessing the properties of atoms.

Below: This computer artwork represents a hydrogen atom—the simplest and most widespread element in the universe. The pink sphere at the center is the proton; wavy lines represent the path of the electron.

Atoms make up all matter—ourselves and everything around us; they are the building blocks of our universe. The ancient Greeks first suggested the idea of these fundamental particles. Back then, however, no one had a method for proving the existence of such tiny particles. Other explanations describing the structure of matter circulated. Thousands of years passed before scientists revived the study of atomic theory at the end of the eighteenth century.

As these scientists returned to the notion of the atom, they faced many questions. How could they prove atoms existed? Were atoms solid? If not, what existed inside them? How did atoms stick together to form everything we see around us? At times it seemed as though the more scientists learned, the more questions developed.

Scientists originally believed that atoms were the smallest particles that could exist. Toward the end of the nineteenth century, researchers discovered even smaller particles inside the atom—protons, neutrons, and electrons. Once scientists realized this, they began

studying how atoms interacted with one another. This knowledge opened up a new field of scientific study known as subatomic physics. Understanding the basic properties of the atom led to the discovery of other chemical elements, the properties of radioactivity, and more. Certain atoms held greater possibilities. In 1938, a team of physicists working in Berlin, Germany, succeeded in splitting the nucleus of a uranium atom. They noticed that the process released enormous amounts of energy. Was there a way to control that energy? Could they use atomic energy to create weapons more powerful than anyone ever imagined? With war looming, the race to harness nuclear energy began.

Knowledge of atomic structure offers more than scientific implications; it also brings practical applications. Electricity, mobile phones, medical techniques, nuclear weapons, and spaceflight all exploit the use of atomic structure. Without knowledge of the internal workings of the atom, no modern industrial society could exist. All this progress occurred in a remarkably short time period. In 1900, atomic research was only important for a few scientists working on it. By 1945, after the United States dropped the first atomic bombs on the Japanese cities of Hiroshima and Nagasaki to end World War II, it became clear that subatomic physics had changed the world—and there was no going back. The story of how the atom was split is an intriguing tale of competition and cooperation, of ingenuity and imagination, of near-miraculous advancements and utter destruction; it is also one of the most dramatic tales in scientific history.

CHAPTER ONE

"Material objects are of two kinds, atoms and compounds of atoms. The atoms themselves cannot be swamped by any force, for they are preserved indefinitely by their absolute solidity."
ROMAN PHILOSOPHER LUCRETIUS (c. 99–55 B.C.)

Understanding Atoms

Although understanding atoms has been a relatively recent phenomenon, the idea of tiny particles making up everything around us was first suggested many centuries ago, in the time of the ancient Greeks. Despite this, it was not until the beginning of the nineteenth century that scientists began to unravel the mysteries of the atom.

Above: *Greek philosopher Democritus was first to suggest that all matter was made up of atoms—tiny, indivisible particles.*

ANCIENT IDEAS ABOUT THE ATOM

As far back as the fifth century B.C., philosophers speculated about the structure of matter and questioned whether everything in the universe consisted of tiny particles—units so small that no further division was possible. They used the word *atomos*, the ancient Greek term for "uncuttable," to describe these units.

Greek philosopher Democritus (*c.* 470–400 B.C.) was one of the first to suggest what we now think of as the modern theory of the atom. Although he did not conduct experiments, make observations, or perform mathematical deductions, he devised several very modern-sounding scientific ideas. One of these was that all matter was made up of solid particles that were

Clockwise from top left: *Water, fire, earth, and air: the four elements that made up all matter, according to Aristotle. Considered a greater philosopher than Democritus, Aristotle's theory of matter was widely accepted, while Democritus's atomic theory was disregarded.*

indivisible; nothing smaller could exist. These particles were so tiny they were invisible to the human eye. While not all of Democritus's suggestions about atoms were correct, many came close to reality.

Unfortunately, there was no way of proving or disproving this theory—people either believed it or they didn't. One of the most famous of the ancient Greeks to disagree with Democritus was the great philosopher Aristotle (384–322 B.C.). Aristotle claimed that a smallest part of matter did not exist, and that all substances were made up of four elements: water, fire, air, and earth. People chose to believe Aristotle's view of the structure of matter. The idea of the atom was not seriously considered again by scientists until the eighteenth century, when new evidence for their existence was discovered.

CHEMICAL ELEMENTS

In 1789, French chemist Antoine Lavoisier (1743–1794) drew up a table of what he called chemical "elements." These elements were substances such as carbon, gold, or oxygen that could not be broken down into simpler substances. Lavoisier identified thirty-three elements and categorized them according to their properties: gases, non-metals, metals, and "earths." Other chemicals are called compounds and are different combinations of the basic elements. Not all of the elements Lavoisier identified appear on the periodic table we use today. Some of

Above: *Antoine Lavoisier drew up a table of chemical elements. It was the first time anyone had attempted to classify substances. Lavoisier's table served as the forerunner of the modern periodic table.*

them were what we now call oxides. They can be divided by chemical reactions. Lavoisier's table was the first time anyone had attempted to classify substances. It paved the way for other scientists to start investigating the elements further.

In 1802, English scientist John Dalton became the first person since ancient times to suggest an "atomic" theory of what the elements were made from. Unlike Democritus, Dalton based his theory on evidence he had gathered by conducting experiments on the chemical elements. Dalton stated that all matter was made of atoms, which he pictured as tiny billiard balls, and which were indivisible and indestructible. These

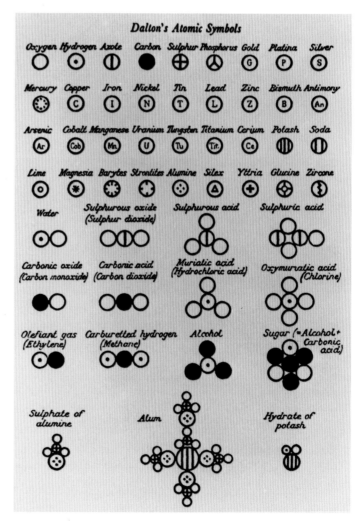

Right: *John Dalton's table of atomic symbols. We now know that some of these are incorrect—water is not HO, for example—but Dalton's ideas about chemical reactions and atoms inspired other scientists to study atomic theories.*

Key People

John Dalton (1766–1844) was an English scientist and teacher. He was fascinated by all types of scientific study and made some important observations about the weather, including what caused rain. His studies of gases led to what we now call Dalton's Law of Partial Pressures, which states that the total pressure exerted by a mixture of gases equals the sum of the pressures of the individual gases. His theory of atoms, however, made him most famous. He thought that all atoms of the same element had the same weight, but atoms of different elements had different weights. He created a table of chemical elements based on their "atomic weights." The modern periodic table uses atomic mass (a measure of matter) as a way of categorizing elements.

Fact

THE IMPORTANCE OF ELECTRICITY

Electricity was a new phenomenon in the early nineteenth century, but scientists quickly harnessed it to use in experiments. This resulted in the discovery of new chemical elements and a greater understanding of the ones that scientists already knew about. Scientist and inventor Humphry Davy (1778–1829) discovered that if he passed an electric current through some substances, they decomposed. (This process is called electrolysis.) When Davy passed an electric current through certain compounds, he found that they separated into their different components. Davy discovered potassium, sodium, and other elements using this method. An electric current passing through water splits it into hydrogen and oxygen in the proportions of two to one (2:1), respectively. This led scientists to speculate that the basic unit of an element was the atom and that all the atoms of a particular element had identical properties.

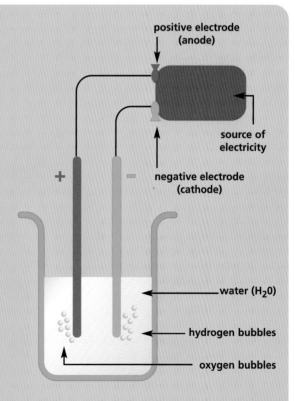

Above: Electrolysis uses an electrical current to break down a substance into its component parts. A current passed through water (H_2O) breaks it down into two parts hydrogen gas and one part oxygen gas.

Above: A "Newton's Cradle" device demonstrates the transfer of energy. As the balls swing, the kinetic energy (energy of motion) transfers from one ball to another until the ball at the other end moves upward. When that ball reaches its highest point—for an instant only—the kinetic energy becomes potential energy (energy of position). As it begins dropping again, the ball's potential energy immediately converts back to kinetic energy.

atoms were made of the pure element. For example, he believed that an atom of gold was made of solid gold.

From that time onward, scientists studied the atomic theory of matter more seriously. In 1815, another English scientist and doctor, William Prout (1785–1850), suggested that perhaps all atoms were simply different arrangements of an even more fundamental "building block." He named this unit the "protyle." Prout thought that a protyle might be like the lightest atom, hydrogen. Different atoms would simply be groups of the same protyle, but in different combinations. Imagine a building site with a pile of bricks: depending on the arrangement of the same amount of bricks you could build a house, a factory, or a railroad station. Similarly, different arrangements of protyles would form different kinds of atoms. Prout's idea of a basic building block influenced later atomic theories—although not in the manner he expected.

ATOMS AND ENERGY

By the middle of the nineteenth century, many scientists were convinced that physical and chemical phenomena, such as heat, light, electricity, and motion, were just different forms of energy that could be converted from one to another.

For example, James Joule (1818–1889) conducted an experiment in which he turned a wheel immersed in a liquid. Some of the kinetic energy (the energy of motion) from the movement of the wheel was converted into heat, or thermal, energy. The temperature of the water increased as a result.

Scientists believed they could better understand this and other experiments if they pictured atoms as all having the same, measurable size. Once scientists defined the size of atoms, they could count them. They imagined that these invisible atoms were extremely tiny spheres—so small that one million atoms would make up the thickness of a sheet of paper.

Left: *Dmitri Mendeleyev was among the first scientists to reorganize the table of elements. He grouped elements that shared similar properties; these groups were called periods.*

Fact

ATOMIC SPECTRA

Heating a chemical until it glows helped reveal the internal structure of that element. For example, sodium produces a characteristic orange color, similar to the color of sodium street lights. Chemists use a glass prism to split the glowing light into its different colors and then map the colors produced by each element. The pattern of colored lights produced by each element is called its spectrum. Researchers studied the spectra of different elements to conclude that all the atoms in a pure element have the same internal structure and produce exactly the same spectrum.

Above: *This is the spectrum of colors from the element helium. Every element has its own unique color spectrum that serves as a kind of "fingerprint" and allows scientists to determine the presence of these elements in other substances.*

Many questions remained. As they learned more about the various forms of energy and the different elements that made up compounds, scientists realized that some kind of force must hold together the atoms in a substance. The incredibly small size of atoms made it very difficult to discover exactly what that force was and how it worked.

As more elements were discovered, several scientists, including German chemist Johann Wolfgang Döbereiner (1780–1849), Englishman John Newlands (1837–1898), and Russian chemist Dmitri Mendeleyev (1834–1907), drew up a table of elements arranged according to similar chemical properties. Mendeleyev's table grouped elements, such as lithium, sodium, and potassium, that shared certain characteristics in one column on his table. This suggested that the atoms in one column or group had structural features in common. But what were these features?

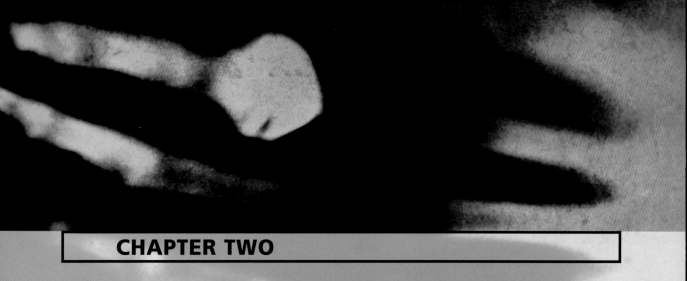

CHAPTER TWO

"Could anything at first sight seem more impractical than a body which is so small that its mass is an insignificant fraction of the mass of an atom of hydrogen?" **PHYSICIST J. J. THOMSON ON THE ELECTRON, 1934**

What's Inside the Atom?

Above: *The first human X-ray, made by Wilhelm Roentgen, showing his wife's hand (wearing a ring). He won the 1901 Nobel Prize in physics for his discovery of X-rays.*

Below: *Roentgen's X-ray machine looks very simple by today's standards. The generator (B) supplied electricity to the cathode-ray tube (T). This generated X-rays, which left an image of the hand on a covered photographic plate (C).*

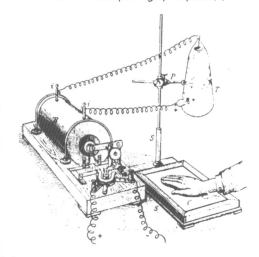

At the beginning of the twentieth century science and technology were developing rapidly. Many people saw the results of this in their everyday lives—electric lights replaced oil lamps, cars replaced horse-drawn carriages. There were also breakthroughs in understanding atoms, and these were built on several key discoveries made at the end of the nineteenth century.

THE DISCOVERY OF X-RAYS

Finding out about X-rays was the first key development that helped scientists on their way to understanding the atom. X-rays were discovered by German scientist Wilhelm Roentgen (1845–1923). During experiments, he noticed a phenomenon that no one had seen before—some unknown force darkened photographic plates that were wrapped up in black paper to protect them from light. Roentgen named the phenomenon X-rays because it was so mysterious. Since X-rays could pass through solid objects, doctors were soon using them to examine broken bones.

Scientists also learned more about atoms by studying X-rays. They are caused by the movement of electrons (the negatively charged particles that move around the nucleus of atoms). In 1913, sixteen years after the electron had been discovered and explained, English physicist Henry Moseley (1887–1915) conducted several experiments with X-rays which revealed that the atomic number in the periodic table represented the positive charge in the atom.

Using a cathode-ray tube, he bombarded different metals with streams of high-energy electrons. He noticed that each of the metals emitted X-rays with different wavelengths. The cathode rays knocked out the innermost electrons from the atoms in the metal, causing electrons in the outer shells to fall into the inner shell. This process emitted X-rays.

The amount of energy needed to knock out an electron from the inner shell of an atom depends on the number of protons in the nucleus. This meant that by measuring the frequency of X-rays emitted by outer electrons falling into the inner shell, scientists could determine the number of protons in the nucleus and the positive charge of an atom.

THE DISCOVERY OF RADIOACTIVITY

Finding one type of radiation, X-rays, prompted a frantic hunt for others. In France in 1896, Henri Becquerel (1852–1908) sorted through a collection of mineral specimens to find the ones that affected photographic plates. He discovered several specimens that blackened the plates in a similar way to X-rays. These mineral samples contained uranium, the heaviest element then known. Becquerel found that uranium naturally emitted a type of radiation that caused changes in objects it encountered.

After Becquerel had discovered it, other scientists began to study radioactive properties in more detail. Polish-born physicist Marie Curie (1867–1934) gave

Right: Henri Becquerel discovered radioactivity in uranium.

Fact

RADIOACTIVITY

When Henri Becquerel and Marie Curie experimented with radio-activity, they did not realize how unusual it was. It was only later that scientists learned of its special properties. Atoms in radioactive elements like uranium are unstable; they emit radiation in the form of particles or rays. Three types of radioactive radiation are alpha particles, beta particles, and gamma rays. Alpha and beta particles cause an element to change, or "decay," into atoms of a different element. The third type of radioactive decay, gamma rays, do not change an element because they are pure energy.

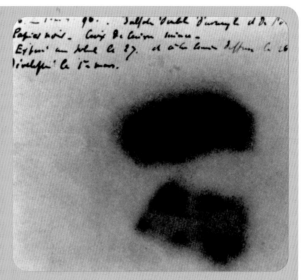

Above: This photograph led to the discovery of radioactivity in 1896. The dark patches show where Henri Becquerel placed crystals of uranium salt on a photographic plate.

the name "radioactivity" to this property. She purified large amounts of uranium ore (pitchblende) to obtain the pure uranium. It was painstaking work, but she was rewarded when, even after removing all the uranium from the ore, it still showed radioactivity. She concluded that other elements in the ore were also radioactive. Curie finally separated small quantities of two previously unidentified elements—radium and polonium.

Both these elements were found to be useful in hospitals for the treatment of cancer, because their radiation killed cancerous cells. The new elements also helped scientists uncover more clues about the structure of the atom.

DISCOVERING THE ELECTRON

By the 1890s, the main use for electricity was for lighting buildings. The new electricity supply companies were in fierce

competition with established companies supplying oil or coal-gas for lighting. The first electric light bulbs were dim, short-lived, and costly. They were equivalent to a modern 25-watt bulb and using them was one thousand times more expensive. Designing better light bulbs became a high priority.

Electric lighting involves passing an electric current through a metal filament in a glass bulb until the filament glows brightly to produce light. The filament must be in a vacuum—the glass bulb must have no air in it at all—or it burns out quickly.

Radio "tubes" were similar to light bulbs; they used metal filaments in glass bulbs. Radio tubes amplified the radio signals. People listened to the transmitted signal through headphones or a speaker. Tubes are the bulky and power-guzzling ancestors of the transistors and integrated

Left: Inventor Lee De Forest (1873–1961), holding an early radio tube. These worked like lightbulbs and used a metal filament within a vacuum in a glass bulb.

circuit boards used in modern radios and other electronic equipment.

In the 1890s, many scientists and engineers studied the behavior of electric current as it traveled through wire in hopes of designing better lighting and radios. One puzzle they encountered was the behavior of electric currents passing through a glass tube that had no air at all inside (a vacuum tube). When the beam of electricity hit the end wall of the tube, it created a glowing spot on the glass. Scientists named the rays that caused this effect "cathode rays" because they originated from the cathode (negative) electrode. These glass tubes are the forerunners of all modern television and computer screens, which are still called cathode ray tubes, or CRTs.

But what exactly were were cathode rays? The debate raged about whether they were particles or waves. Working in Cambridge, England, in 1897, physicist J. J. Thomson (1856–1940) finally discovered the answer. In his experiments, Thomson found he was able to change the direction of a beam

Below: This type of cathode-ray tube is called a "Crookes' tube." British physicist Sir William Crookes used such a tube in 1878 to investigate cathode rays. Crookes's experiments led J. J. Thomson to discover that cathode rays were actually beams of electrons.

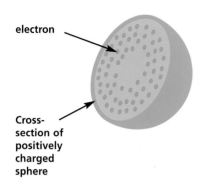

electron

Cross-
section of
positively
charged
sphere

*Above: Thomson used his discovery
of the negatively charged particles
(electrons) to design a simple
representation of the atom that
became known as the "plum-pudding"
model. In it, the blue dots—the
electrons—were arranged like
currants inside a plum pudding.*

of cathode rays using both magnetic and electric fields. Thomson's experiments proved that cathode rays were particles with a negative electric charge. He also suggested their mass was only 0.0005 that of a hydrogen atom.

He called his new particle the "corpuscle," but the name didn't catch on and it was renamed the "electron." This unit of negative charge was the first subatomic particle discovered.

The discovery of the electron raised even more questions. Electrons are negatively charged. Atoms are electrically neutral. This meant that inside the atom there must be a positive charge that cancelled out the negative charge. What was it? And what other secrets did atoms have locked away?

FINDING THE NUCLEUS

Encouraged by these discoveries, physicists conducted more experiments. They noticed that polonium atoms—one of the elements discovered by

Fact

ELECTRONS—WAVES
OR PARTICLES?

J. J. Thomson believed
the electron was a
particle because it
behaved like other
particles. As quantum
mechanics—the study of
wave phenomena at the
particle level—developed
in the 1920s, it became
clear that electrons
could also behave like
waves. For example,
electrons produce
interference patterns
similar to those made by
light waves. G. P. Thomson
(J. J. Thomson's son)
demonstrated the wave
properties of electrons
in 1927.

*Above: Light behaves like a wave because it produces
interference patterns like these. Pebbles dropped into water
cause tiny waves that move outward in increasing circles.
The waves caused by one pebble eventually meet and interfere
with the waves of other pebbles.*

Fact

THE PHOTOELECTRIC EFFECT

In the early twentieth century, scientists suggested many theories about the processes they observed in chemistry and physics. Albert Einstein's (1879–1955) 1905 theory of the photoelectric effect proved to be one of the most important theories of all.

Einstein puzzled over the fact that only certain colors, or wavelengths, of light falling on a metal plate ejected electrons from the plate. He theorized that the light hit the metal in a series of "packets" that he called "quanta." The energy of these quanta depended on the color of the light: blue or ultraviolet light had high energy, while red light had low energy. Only packets with a certain level of energy were absorbed by the atoms in the plate, causing the ejection of electrons. Einstein's theory of the photoelectric effect made other scientists think about light as traveling in a series of packets as well as in waves.

The photoelectric effect is still important today in devices such as video cameras and televisions. It can convert light from an image into electrical signals for recording or broadcasting. The theory also explains how light stimulates the production of glucose in green plants. We now know that the photoelectric effect is the starting point for all food chains.

Right: *Albert Einstein was awarded the Nobel Prize for physics in 1921 for his explanation of the photoelectric effect.*

Marie Curie—self-destructed in a process called radioactive decay. During radioactive decay, the atoms disintegrate by firing off a large chunk (an alpha particle). Alpha particles were the ideal probes to find more information about the structure of the atom. Alpha particles were later proved to be helium atoms minus their electrons.

In Manchester, England, Hans Geiger (1882–1945) and Ernest Marsden (1889–1970) conducted an experiment in which they aimed these alpha particles at gold foil—extremely thin sheets of gold. Most alpha particles passed straight through the gold foil, but unexpectedly a very small number hit the foil and bounced back. Geiger and Marsden could not explain why.

Below: Most alpha particles fired at a thin sheet of gold foil pass straight through it. If an alpha particle directly hits the nucleus of one of the gold atoms, the positive charge of the nucleus repels the positively charged alpha particle and bounces it back.

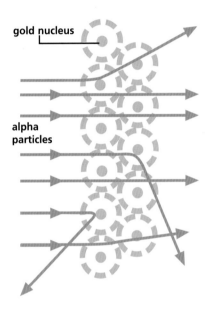

Below: Rutherford's model of the atom. The nucleus contains more than 99.9 percent of an atom's mass. The nucleus carries a positive charge that holds the negatively charged electrons in orbit.

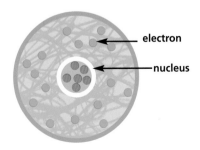

In 1911, English physicist Ernest Rutherford came up with the explanation for this: Atoms in the gold foil must be made up of mostly empty space that allowed the alpha particles to pass straight through. The interesting phenomenon was that a few of the particles bounced back as if they had hit something solid. Rutherford suggested that each atom contained a central core, or nucleus. He believed the nucleus was only about 0.0001 the diameter of the atom, but that it contained most of the mass of the atom—as well as the positive charge that scientists had been looking for. The positive charge of the alpha particles caused them to bounce back when they collided with another positive charge, so the nuclei of the atoms in the gold foil must also be positive. Rutherford compared the collisions between alpha particles and the nuclei to battleship shells hitting a piece of tissue paper—and rebounding!

THE PLANETARY MODEL OF THE ATOM

Danish scientist Niels Bohr (1885–1962), took the next step. Two years after Rutherford explained his theory of the atom, Bohr came up with a different model, based on the same idea. He agreed that the electrons orbited the nucleus, but he suggested that these orbits took the form of a number of fixed "shells," with the electrons moving around in them. This became known as the planetary model of the atom. A shell, or orbit, could contain one or two electrons whizzing around. Bohr said that electrons in the orbits closest to the nucleus had less energy than those in orbits farther away. While it stayed in a particular orbit, the energy of the electron was fixed. When an electron moved closer to the nucleus into a lower energy orbit, it emitted the excess energy as a flash of light.

Bohr's electron shell theory explained the particular colors of atomic spectra. Specific wavelengths, or frequencies, of light—and only those—appeared or disappeared when an electron moved from one energy level, or orbit, to another.

Key People

Ernest Rutherford (1871–1937) was one of the most talented physicists of the early twentieth century. After receiving the 1908 Nobel Prize in chemistry for his study of radioactive elements, he began studying the nature of the atom. In 1911, he explained the famous gold-foil experiment which showed that there was a positively charged nucleus at the center of an atom. He became director of the famous Cavendish Laboratory in Cambridge, England (a post previously held by J. J. Thomson), in 1919. During his tenure there, he made many contributions to physics, including researching nuclear reactions. Rutherford did not believe the energy produced in these reactions could ever be harnessed by scientists. It turned out to be his greatest mistake. His worked helped pave the way for the discovery of nuclear fission (*see* p. 27) and the development of nuclear weapons.

Below: Bohr's atomic model showed electrons moving around the nucleus in orbits or "shells," in much the same way as the planets in our solar system move around the Sun. Although his model, known as the planetary model of the atom, was not quite complete, it helped scientists come closer than ever before to an accurate understanding of atomic structure.

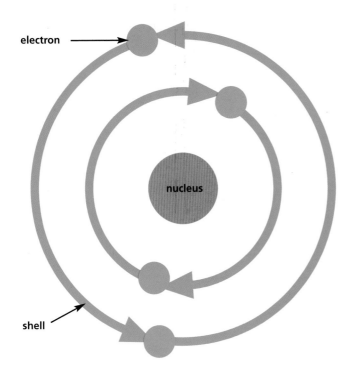

electron

nucleus

shell

Fact

ELECTRON SHELLS

The maximum number of energy levels, or shells, in an atom is seven, and each shell holds an increasing number of electrons. The first shell (the one closest to the nucleus) holds only two electrons; the second holds up to eight; the third up to eighteen. Theoretically, the seventh shell holds up to ninety-eight electrons.

PROVING THE EXISTENCE OF THE PROTON

Scientists now knew that the nucleus had a positive electric charge, but they still didn't know what caused it. Did the negatively charged electrons have positively charged counterparts? If so, where were they? Another English scientist, Francis W. Aston (1877–1945), took a step toward answering this question. He was working on experiments that would "weigh" atoms very precisely, and he managed to measure the masses of neon atoms. He thought that all atoms of the same element would have the same mass, but to his surprise he found two different measurements of mass in neon atoms. These are two different varieties, or isotopes, of neon. One isotope is twenty times, the other twenty-two times the mass of hydrogen. Aston went on to invent a device called a mass spectrograph, which allowed him to separate the isotopes of other elements. He used this invention on other chemical elements and discovered 212 other isotopes. In 1922, Aston was awarded the Nobel Prize in chemistry for his work.

Right: Aston's mass spectrograph separated isotopes of various elements. The globe contained a compound of the substance being tested. An electric current passing through the globe knocked electrons from the atoms of the substance.

Rutherford was interested in Aston's discovery of isotopes because it reminded him of William Prout's theory of protyles: perhaps the hydrogen nucleus was a building block of all nuclei after all. Rutherford suggested the name "proton" for this unit.

In 1925, Patrick Blackett (1897–1974), a young research student at the Cavendish Laboratory provided further evidence for the existence of the proton in the nucleus of an atom. He successfully photographed a collision between an alpha particle and a nitrogen nucleus. A positively charged particle—a proton—emerged from the nitrogen nucleus after the collision.

Above: Blackett's photograph shows alpha particles scattering from atomic nuclei of different masses. Here, the alpha particles stream upward through the cloud chamber; one of them scatters a proton (the fainter track moving from bottom right to upper left) through the water vapor that fills the chamber. The heavier alpha particle is slightly deflected to the right.

Fact

CLOUD CHAMBERS AND GEIGER COUNTERS

Atoms and subatomic particles are so small that they are impossible to see even with the best optical microscopes. In the early twentieth century, new methods were devised to detect subatomic particles. Blackett used one of these new methods—the cloud chamber, a cylinder filled with water vapor—when he photographed the collision between the alpha particle and the nitrogen nucleus. When an electrically charged particle passed through the chamber, it knocked off some of the electrons. Water droplets formed on them. A trail of droplets traced the track of the particle through the cloud chamber. These were photographed and analyzed.

Counting large numbers of particles with a cloud chamber proved difficult, however. Early versions of the Geiger-Muller counter detected charged particles as they entered the counter tube and triggered an electric current. The the spurts of current measured the number of particles passing through the counter.

29, 1932

A NEW RAY

DR. CHADWICK'S SEARCH FOR " NEUTRONS "

Dr. James Chadwick, F.R.S., Fellow of Gonville and Caius College, Cambridge, in an interview on Saturday said that the result of his experiments in search of the particles called " neutrons " had not, at the moment, led to anything definite, and the element of doubt about the discovery still existed.

There was, however, a distinct possibility that investigations were proceeding along the right lines. In that case a definite conclusion might be arrived at in a few days, and on the other hand it might be months. Dr. Chadwick described his experiments as the normal and logical conclusion of the investigations of Lord Rutherford 10 years ago. Positive results in the search for " neutrons " would add considerably to the existing knowledge on the subject of the construction of matter, and as such would be of the greatest interest to science, but, to humanity in general the ultimate success or otherwise of the experiments that were being carried out in this direction would make no difference.

Lord Rutherford, at the conclusion of his lecture at the Royal Institution on Saturday, confirmed in a statement to a representative of The Times the importance of the experiments by Dr. Chadwick. They seemed to point, he said, to the existence of a ray whose particles, known as " neutrons," were indifferent to the strongest electrical and magnetic forces. He did not, however, confirm the assumption that these particles were matter in the everyday sense of the word from the fact that their collisions obeyed the laws of momentum ; nor the further assumption that they moved at a speed more than a tenth of that of light or electricity.

Lord Rutherford's own lecture dealt with the " Discovery and Properties of the Electron," and in the course of it he conducted a number of experiments in the generation of cathode rays, with glass tubes and bulbs used 50 or more years ago by Sir William Crookes, and with others lent by Sir William Pope.

The discovery in 1897, said Lord Rutherford, of the negative electron had been of profound significance to science. The electron tube was essential to-day not only for the generation of continuous radiations, but for their reception, and thus rendered possible the rapid development of radio-telephony and broadcasting. When they looked back, it became clear that the year 1895 marked what they might call the definite line between the old and the new physics. It was in that year that Röntgen made his famous discovery of the X-rays. The importance of the discovery was really, in a sense, greater than itself, because it gave the impetus to experiments which led to two epoch-marking discoveries, of radioactivity by Becquerel in 1896 and of the electron in 1897. Those two discoveries opened up new vistas of the wonderful ways in which Nature worked. They gave us, for the first time, methods for attacking the question of the structure of the atom, and we now had a fairly definite general view of that structure; and they had given us, for the first time, a fairly clear idea of the mechanism of radiation.

Above: In 1932, the world learned of Chadwick's discovery of the neutron, the particle that existed along with protons in the nucleus of an atom.

WHAT WERE THE MISSING PARTICLES?

There was still a missing link. In order for atoms to be neutral, the positive charge in the nucleus must equal the negative charge so that they cancel each other out. Experiments that measured the weight of atomic nuclei showed a discrepancy. For example, a nitrogen atom nucleus weighs as much as fourteen protons—but it only has seven electrons. How could the atom be electrically neutral if it had more protons than electrons? Since scientists only knew about these two particles, some of them suggested that the atom contained additional electrons—in the nucleus—which neutralized some of the extra proton charges. If this were the case, scientists believed that a nitrogen atom would have fourteen protons and seven electrons in the nucleus, and a further seven electrons in the shells outside the nucleus.

Many scientists did not support the idea that extra nucleic electrons answered the puzzle, though, and researchers in laboratories around the world joined the challenge of finding the missing part of the nucleus. James Chadwick discovered the answer in 1932—once again in Cambridge, England. He conducted experiments in which he bombarded atoms of beryllium with alpha particles. What he found was another constituent of the nucleus, a particle with the same mass as the proton but no electric charge. Scientists called this the neutron. That meant nuclei contained protons and neutrons: Nitrogen has seven protons and seven neutrons in its nucleus.

Finally, scientists had cracked the atom—they knew it contained protons, neutrons, and electrons. As it turned out, this knowledge was just the beginning. Atomic structure was not at all simple—an atom of uranium, the heaviest known element, was a complex structure containing hundreds of protons and electrons. Research did not stop there. Physicists began to learn more and more about the nature of atoms and what they could do.

Left: Beryllium—the element James Chadwick used in his experiments to reveal the neutron—has four electrons orbiting in shells outside the nucleus. Beryllium's nucleus (the pink sphere in the middle), contains four protons and—as Chadwick discovered—five neutrons.

Key People

James Chadwick (1891–1974) found the missing piece of the puzzle of atomic structure when he proved the existence of the neutron—the particle in the nucleus of an atom with no electrical charge. Chadwick began his studies on particle behavior under Ernest Rutherford at the Physical Laboratory in Manchester, England, in 1911. He also spent some time working with Hans Geiger in Germany, but eventually returned to England, where he rejoined Rutherford, now working at the Cavendish Laboratory in Cambridge, England.

His breakthrough came in 1932 after he began experiments in which he bombarded atoms of elements with alpha particles. These experiments revealed the long-sought-after neutral particle that explained the discrepancy between an element's atomic number and its atomic mass. Chadwick was awarded the Nobel Prize for physics in 1935 and was knighted in 1945.

By using streams of particles in this way, Chadwick helped pave the way for developments in nuclear fission that occurred rapidly over the following fifteen years, culminating with the detonation of the first atomic bombs in 1945.

CHAPTER THREE

"If I could remember the names of all these particles, I'd be a botanist." PHYSICIST ENRICO FERMI

Unlocking the Secrets of the Nucleus

As the twentieth century progressed, new technology became available that allowed scientists to delve even deeper into the structure of the atom and to understand how the subatomic particles worked. Most important, they wanted to crack open the nucleus. Once they knew how to do this, they could manipulate atomic structure. The study of the nucleus became known as nuclear physics, and it set scientists on the path to creating a form of power greater than anything mankind had ever known.

Above: A colorized image shows the emission of radioactive alpha particles from the element radium.

ACCELERATORS AND COUNTERS

Until the 1930s, scientists and researchers used radioactive materials, such as radium or polonium, as source material for the fast-moving particles needed to bombard nuclei. These radioactive sources were difficult to prepare, expensive to obtain, and dangerous to use. Researchers wondered if there was an alternative substance that could provide the necessary particles in a more efficient way. They also began thinking about how to create a machine that could send a beam of particles traveling at high speeds to facilitate these experiments.

Left: A modern particle accelerator, based on Cockcroft and Walton's original, is used in the first stages of particle acceleration at Fermilab.

Developments in the electrical industries also helped research technology catch up with scientific theory around this time. Researchers now had access to large magnets for use as electromagnets, transformers for producing high voltages of electricity, and air pumps for creating efficient vacuums inside chambers and tubes. With this better equipment at their disposal, teams of physicists and engineers raced to build the first particle accelerator, or "atom smasher."

John Cockcroft (1897–1967) and Ernest Walton (1903–1995) built the first working accelerator in the Cavendish Laboratory. It used 100,000 volts of electricity to accelerate the flow of protons. Was that high enough to force protons into the nucleus of a target atom? Remarkably, it was.

Earlier calculations showed that alpha particles could "leak" out of the nucleus. Cockcroft reasoned that it should also be possible for protons to burrow back in. His theory proved correct. In 1932, Cockcroft and Walton became the first people to split the atom. Proton "bullets" shot at lithium atoms made the nuclei disintegrate and produced energy. Einstein's $E=mc^2$ equation worked!

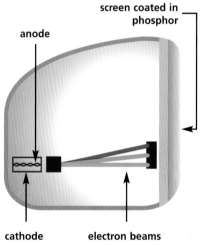

Above: An ordinary television or computer screen works much like a miniature particle accelerator. The alignment of the cathode and anode form a sort of "electron gun" that speeds up electron movement and smashes them into a phosphor-coated viewing screen. The collision causes a "pixel" of color on the screen. Particle accelerators work in a similar way, except that the particles move much faster and the collision results in lots of subatomic particles.

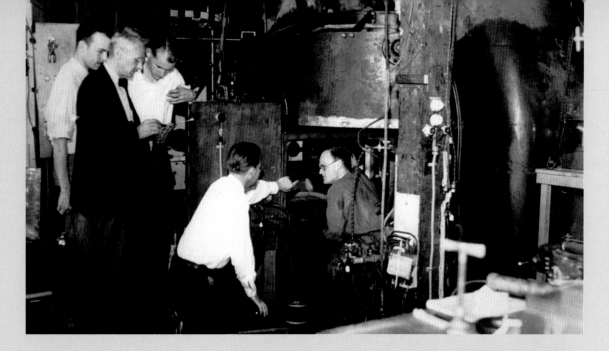

Above: Ernest Lawrence (pointing) demonstrates the cyclotron.

Within weeks, Ernest Lawrence (1901–1958) and Stanley Livingston (1905–1986) in Berkeley, California, had similar success. They called their particle accelerator the cyclotron. Particles in the cyclotron traveled in circles at ever-increasing speeds. Today, most of the large accelerators are based on a design similar to the cyclotron.

The work of Lawrence at the Radiation Laboratory in Berkeley paved the way for what became known as "Big Physics"—experiments involving large teams of physicists and engineers, huge laboratories, and enormous budgets.

WHAT DID PARTICLE ACCELERATORS REVEAL?

Using particle accelerators for experiments in nuclear physics had one significant advantage: Physicists knew the kinetic energy of the protons. Using protons with known energy to bombard nuclei in a target meant that scientists could accurately calculate the energy involved in any nuclear reaction. For the first time, Einstein's famous equation linking energy and mass, $E = mc^2$, could be measured.

Using the latest technology, German physicist Werner Heisenberg (1901–1976) showed that a specific force, the strong interaction force, held together atomic nuclei. The other fundamental forces included the gravitational and electromagnetic forces that scientists already knew about.

One big puzzle remained. In one type of radioactivity—beta decay—an electron was emitted from the nucleus. To many scientists, this seemed to back up the idea that electrons did in fact exist in the nucleus—as they had believed before the existence of the neutron was revealed. As before, they thought these "nuclear" electrons were in addition to the electrons orbiting in shells around the atom. The problem was that this theory contradicted recent research.

In 1934, Italian physicist Enrico Fermi suggested a new explanation for beta decay. He explained that at a point in the radioactive decay, a neutron actually changed into a proton. In the same instant, an electron was created. Fermi also suggested that in order to ensure energy was conserved, a neutral particle with very little mass was emitted alongside the electron. He called this particle the "neutrino" (Italian for "little neutral one") to distinguish it from the neutron.

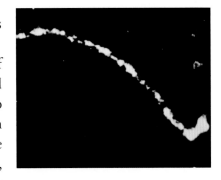

Above: *A photograph of beta decay of a helium nucleus. The short, thick track at the bottom right is the nucleus; the curving track moving away from it is an electron. The electron was created during the decay process.*

NUCLEAR FISSION

What other secrets remained locked inside the nucleus? Over the next few years, scientists all over the world tried to find out. They soon realized that instead of using protons, which are positively

Key People

Enrico Fermi (1901–1954) was an Italian physicist. His first great success came in 1934 when he solved the mystery of beta decay. He continued his work on radioactive elements and soon proved that the nucleus of almost any element could be transformed if it was bombarded with neutrons, rather than protons as had been previously used. His experiments with neutron bombardment led to the discovery of nuclear fission and a turning point in the history of understanding the atom. Fermi later became a United States citizen. Fermilab is named after him. Enrico Fermi was awarded the Nobel Prize for physics in 1938.

Fact

$E=mc^2$

One of Einstein's greatest theories was that matter and energy are just different forms of the same thing. Matter can therefore be turned into energy and vice versa. In Einstein's famous equation, $E=mc^2$, E is energy, m is mass and c^2 is the speed of light squared. To find the energy, multiply the mass by the speed of light squared. When a stream of neutrons collides with a nucleus, it forces out some of the mass of that nucleus, releasing it in the form of energy. Scientists measured the energy released during experiments to confirm Einstein's equation.

Right: Enrico Fermi (on left) and his team in Italy worked tirelessly to solve the "Uranium Problem" of why bombardment with neutrons created many elements with unexpected levels of radioactivity. German scientists Otto Hahn, Fritz Strassmann, and Lise Meitner found the solution to the Uranium Problem in 1938.

charged, as the "smashers" in their experiments, it made more sense to use neutrons. Unlike protons, neutrons were not repelled by the positive electric charges in the nucleus when they got near it and could therefore be much more effective.

Fermi pioneered this technique at his laboratory in Rome, Italy. His team found a way of producing a stream of neutrons and then bombarded elements such as uranium with the particles. Through this process, Fermi hoped to create a few atoms of some elements that did not occur naturally on Earth. These created elements would be even heavier than uranium—the heaviest known naturally occurring chemical element.

In their attempt to create the first man-made, or "artificial" elements, Fermi and his team looked for sources of hydrogen atoms. They even used the water in the goldfish pond on the grounds of their

laboratory. Members of the team in Rome were rather confused by the results of their experiments with neutrons and uranium. They found that many different types of radioactivity were produced, but they could not explain why. They called this the Uranium Problem.

News of the Uranium Problem reached two chemists, Otto Hahn (1879–1968) and Fritz Strassmann (1902–1980), working in Berlin, Germany. Together with Lise Meitner (1878–1968), they solved the problem in 1938. They realized that by bombarding uranium with neutrons, they actually caused the atom—or more specifically, the nucleus of the atom—to split into smaller pieces, creating two different radioactive elements. They had, quite literally, split the atom.

They named this process "nuclear fission" and pointed out that a fission reaction released huge amounts of energy—much more energy than that released by other atomic or nuclear processes.

Above: Lise Meitner and Otto Hahn, whose experiments with neutron bombardment of uranium revealed that the nucleus of an atom could be split to create other elements with lower atomic numbers.

Fact

ANTIMATTER

At around this time, theoretical physicist Paul Dirac (1902-1984) made a surprising prediction in Cambridge, England. He predicted the existence of antimatter. What he meant was that scientists at the time knew atoms and particles existed in a state of positive energy, so according to physics, there must also be negative energy. Dirac believed that for every particle that exists, there is an antiparticle. Since particles consist of matter, antiparticles must consist of antimatter. They have exactly the same mass as the particle, but an opposite electric charge. American physicist Carl Anderson (1905-1991) soon proved Dirac's prediction. Anderson studied cosmic rays (high-energy particles from space). Among them he also found positive electrons, or positrons. We now know that every particle has an antimatter equivalent.

Above: A particle accelerator causes a burst of energy (long green line) that produces particles of opposite charges: a negative electron (green spiral) and a positron (red spiral).

CHAPTER FOUR

"The energy produced by the breaking down of the atom is a very poor kind of thing. Anyone who expects a source of power from the transformation of the atom is talking moonshine."

PHYSICIST ERNEST RUTHERFORD, 1933

Fission Unleashed

Above: *Scientists at the University of Chicago, Illinois, witness the first controlled nuclear chain reaction (in reactor, at right).*

Below: *A page from Fermi's patent for what he called the "neutronic reactor." It was based on the first nuclear pile reactor, the blueprint for the nuclear reactors that followed.*

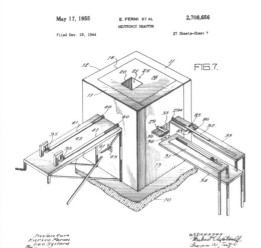

The discovery of fission in the winter of 1938–1939 opened up a new path for nuclear scientists. If nuclear fission could be harnessed and controlled, the energy released could be very powerful. As World War II loomed, physicists in several countries speculated that fission might be used to make a very powerful bomb.

SETTING OFF A CHAIN REACTION

When one nucleus of uranium is split, the energy released is the equivalent to that of a flea jumping. The fission of one nucleus—one flea jump—is useless. Scientists aimed to get more "fleas" to jump together and use that combined energy. The uranium fission process releases neutrons. Physicists were intrigued with the possibility that the neutrons released during one uranium fission could "pile up" to make other uranium nuclei fission—releasing more neutrons that could fission other uranium nuclei, and so on. This process is called a nuclear chain reaction, and it releases enormous amounts of energy.

If the chain reaction went slowly, the release of the energy would be steady and under control. As long as it was controlled, uranium could be used as a fuel in a reactor to produce heat for generating electricity—a nuclear power plant. If the chain reaction went very fast, almost instantaneously, then the amount of energy released would be so great and uncontrollable that it would cause a huge explosion: an atomic bomb blast.

Only one isotope of uranium, uranium-235 (U-235), was efficient enough to be used in nuclear fission. Uranium-235 made up less than one percent of naturally occurring uranium, and its nuclei were too rare to be hit by passing neutrons. The more common, naturally occurring element, uranium-238 (U-238), absorbed neutrons, making it very difficult—if not impossible—to set up a chain reaction from ordinary uranium. The physicists' initial excitement wore off.

THE ROAD TO NUCLEAR WEAPONS

World War II broke out in 1939. The Germans believed that the war would be over very quickly, so even if nuclear weapons were possible, they would not be developed in time to affect the outcome of the war. They regarded nuclear research as a longer-term study (at least at the beginning of the war), and did not make it a high priority.

The Allies worried that the Germans might develop a nuclear weapon. They gathered together their best physicists and invested a lot of money to find a way to cause a fission reaction in ordinary uranium.

Refugee physicists from Germany, Otto Frisch and Rudolf Peierls, worked to separate the isotope U-235 from U-238, to create a nuclear weapon. This work led to the Tube Alloys Project in Britain, which later merged with the much larger Manhattan Project in the United States.

Left: *The Los Alamos laboratory in New Mexico was the heart of the top-secret Manhattan Project that developed atomic weapons using nuclear fission.*

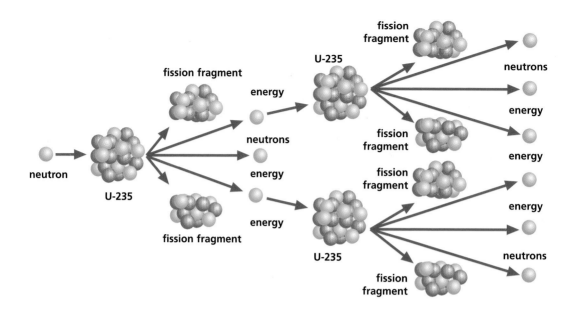

Above: *During a nuclear chain reaction, neutrons fired at an element, such as U-235, collide with the nucleus, splitting it into fragments and releasing energy and neutrons. These neutrons, in turn, collide with other uranium nuclei, creating more fragments and releasing additional neutrons and energy.*

The Manhattan Project was a top-secret government project involving scientists and physicists from around the world with the express purpose of creating an atomic bomb from radioactive elements, such as uranium. Separating the fissionable U-235 isotope from the more common U-238 proved to be a difficult, physical process. While the scientists investigated the use of uranium isotopes for the bomb, they also pursued other possibilities.

Key People

Otto Frisch (1904–1979) and **Rudolf Peierls** (1907–1995) were the two men who developed a method to isolate enough U-235 to use in a nuclear weapon. Frisch was born in Austria and Peierls was German, but both men fled to England at the beginning of World War II and helped the British with atomic research. They were both key members of the team brought together for the Manhattan Project in the United States. This project finally developed a successful atomic weapon. After the war, both men returned to England. Frisch worked at the famous Atomic Energy Research Establishment at Harwell, England. Peierls became a professor of physics at Birmingham University and Oxford University.

The most promising of these was to explore the fissionable properties of a completely new isotope of plutonium. A nuclear chain reaction converted some of the U-238 neutrons into plutonium-239 (Pu-239), a fissionable material. Plutonium could then be separated from the uranium by chemical methods.

In December 1942, Fermi and his colleagues built the first nuclear reactor (also called a "pile" reactor) in Illinois and set off a nuclear reaction. Once they proved it worked, the reactor was dismantled and larger reactors were built for plutonium production in Hanford, Washington.

The purified U-235 and the Pu-239 were incorporated into weapons at Los Alamos, New Mexico. On July 16, 1945, the new plutonium weapon was successfully tested near Alamogordo, New Mexico. Within one month, two atomic bombs were dropped on the Japanese cities of Hiroshima and Nagasaki. Hundreds of thousands of people died, including American prisoners of war. Soon afterward, the Japanese surrendered and WWII was over.

The sheer power of these bombs and the devastation they caused shocked the world. Just about one century after John Dalton had revived the atomic theory of matter, scientists had used the atom to create a weapon that changed the world.

Above: *The nuclear devastation of Hiroshima, Japan, shocked the world and helped end World War II. For the first time, people realized the terrifying power of nuclear weapons.*

Background: *The atomic explosion after the detonation of the first atomic bomb test at Alamogordo, New Mexico, in July 1945.*

Fact

URANIUM ISOTOPES
All the atoms of uranium have 92 protons in the nucleus—this is what defines them as uranium. The number of neutrons in uranium nuclei differs, however. The most common isotope, uranium-238, has 238 protons and neutrons. That total includes 92 protons, so the number of neutrons is 238 minus 92 = 146. The number of neutrons in uranium-235 is 235 minus 92 = 143 neutrons.

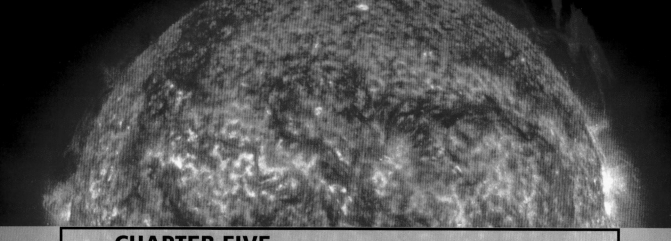

CHAPTER FIVE

"The unleashed power of the atom has changed everything save our modes of thinking and we thus drift toward unparalleled catastrophe." **PHYSICIST ALBERT EINSTEIN, 1946**

A Nuclear Future?

After World War II, several countries—including Britain, Russia, and China—set up programs to develop nuclear weapons even further. Making plutonium required nuclear reactors. The heat from these reactors could produce steam that could generate electricity. Gradually, nuclear reactor designs developed for the military were used in civilian nuclear power plants.

Above: *Nuclear fusion takes place naturally in the Sun, creating enough energy to keep it burning. Hydrogen nuclei fuse to create helium nuclei.*

NUCLEAR FUSION

Nuclear fusion is the opposite of fission. Instead of splitting the nuclei, the nuclei fuse (join together). This is what happens in the Sun. Energy is released during nuclear fusion just as it is during nuclear fission. Machines built to cause nuclear fusion are called fusion reactors.

Scientists and engineers have tried to use deuterium, an isotope of hydrogen, in a fusion reactor. The technology required is very advanced and still has not been completed. The most promising design for a fusion reactor is the tokamak, a doughnut-shaped device developed by the Russians.

To achieve fusion, deuterium must be in the form of a plasma, a state in which the nuclei have been

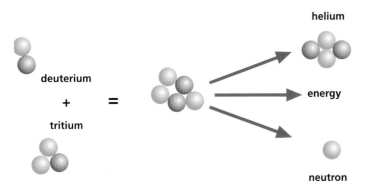

deuterium

+

=

tritium

helium

energy

neutron

Left: In nuclear fusion, two hydrogen isotopes (in this case deuterium and tritium) are fused. In the process, energy is released in the form of a neutron. The reaction creates helium.

stripped of their electrons. To create fusion, the plasma is heated to very high temperatures. It must also be kept away from the walls of the tokamak, so it is confined by precisely shaped magnetic fields.

Building a reactor that produces more energy than it consumes has proved more difficult than many of the early supporters of fusion research expected.

Below: This huge machine is the Joint European Torus (JET) at Oxfordshire, England. The ring-shaped chamber confines plasma that is used in fusion reactions. The machine helps scientists investigate nuclear fusion and search for ways to use it as a source of commercial power.

The most recent international project is JET (Joint European Torus), built in England. This has provided information that will be used to design a bigger and better machine, the ITER (International Thermonuclear Experimental Reactor). The ITER is one step closer to a design for a commercial nuclear fusion power plant.

Fusion as a source of commercial power is still some decades away; however, if it can be made to work safely and efficiently, fusion will provide cheap and readily available power. The machines themselves will become highly radioactive and will have to be maintained by robots rather than human. Unlike present-day nuclear power plants, fusion plants will not produce radioactive waste, so disposing of hazardous wastes is not an issue.

Below: *The explosion caused by the detonation of the first hydrogen (or thermonuclear) bomb. The United States set it off on Eniwetok atoll in the Marshall Islands in the South Pacific on November 1, 1952.*

Fact

ELECTRICITY FROM NUCLEAR POWER

The first nuclear power plant in Britain, Calder Hall, opened in 1956. This originally produced plutonium for weapons, but was soon supplying electricity for the National Grid. The United States developed a compact reactor—the Pressurized Water Reactor (PWR)—designed for propelling nuclear submarines, but it was first used to generate electricity on land at Shippingport, Pennsylvania.

Right: Nuclear power plants like this one in Switzerland generate power for many different purposes—from creating the radioactive elements used in weapons to generating electricity for commercial use.

NUCLEAR WEAPONS

During the development of the atomic bomb based on the fission process, several scientists speculated that it might be possible to build an even more powerful weapon based on the fusion process. These weapons are called hydrogen, or thermonuclear, bombs.

Hydrogen bombs are actually detonated by atomic (fission) bombs. This explosion creates the conditions for the fusion process to take place. Fusion weapons can be built much larger than fission weapons and are much more destructive.

A number of countries around the world possess the capability of producing nuclear weapons. International treaties help reduce nuclear arsenals and prevent the spread of weapons to non-nuclear countries. These negotiations have had significant but limited success. One major concern about the spread of civilian nuclear power plants is that it gives countries access to nuclear technology, some of which can be used to develop nuclear weapons of mass destruction.

Below: The United States dropped a fission bomb, nicknamed "Little Boy," on Hiroshima in 1945. Scientists began thinking of ways to produce even more powerful nuclear weapons—fusion, or thermonuclear bombs—almost as soon as fission bombs were created.

CHAPTER SIX

"The terror of the atom age is not the violence of the new power but the speed of man's adjustment to it—the speed of his acceptance."

AMERICAN AUTHOR E. B. WHITE

Particle Physics Today

Above: Atomic particles subjected to a high-speed collision form colored tracks in a bubble chamber. (see p. 39.) The tracks allow scientists to study particle behavior.

The apparent simplicity of the early days of nuclear physics came to an abrupt end. The model of the atom in which a cloud of electrons surrounded a nucleus consisting of protons and neutrons could not explain a spate of new discoveries: the positive electron, the neutrino, strange particles . . . the list grew longer and longer.

TECHNOLOGY SINCE WORLD WAR II

The model became complicated in the 1950s and 1960s with the development of a new generation of much larger and more powerful accelerators. In part, this became possible because of new technologies—such as radar—developed during World War II. Also, physicists enjoyed a high postwar reputation because of their achievements. This meant governments were more willing to fund their research.

The revolution in accelerator design led to more efficient particle detectors. These were developed more through necessity than anything else. Scientists needed accurate methods to capture and analyze the results of experiments. Simply keeping up with the flood of research data generated by the new machines proved challenging.

One of these new machines was the bubble chamber. Broadly based on the cloud chambers that helped researchers identify radioactive particles before World War II, bubble chambers are much more sophisticated technology. They use liquid hydrogen, a very dense medium, to study the behavior of atomic particles. As the beam particles smash into the hydrogen nuclei, they leave trails of bubbles throughout the liquid hydrogen. These trails are

Above: The particle accelerator at CERN, the European center for particle physics near Geneva, Switzerland. In 1932, James Chadwick used one of the first particle accelerators to discover the neutron.

Fact

THE BIG BANG

Accelerators like the ones at Fermilab and CERN probe the structure of matter in minute detail. In accelerator reactions, conditions begin to approach those that scientists believe existed soon after the Big Bang—the point at which the universe was created. Accelerator experiments not only provide information about the properties of particles, but also shed light on what happened in the fractions of a second after the Big Bang billions of years ago.

Above: Fermilab. The figure-eight shape shows the two main particle accelerators. These devices revealed information about atomic structure that was unattainable only fifty years ago.

photographed, recorded, and measured to give details of the particles involved in any interactions that occur. When bubble chambers were first used, the task of analyzing the tracks was immense—tens of thousands of photographs had to be scanned. The invention of faster and more powerful computers meant that new kinds of electrical detectors could be built that selected specific particle "events" for scanning. An "event" is a collision of two particles or the decay of one particle. Computerized information about these events make nearly instantaneous analyzation possible and allow researchers to conduct larger numbers of experiments as they look for rare particle interactions.

Today, research on particle physics is carried out in large international laboratories with huge accelerators—the larger and more powerful descendants of the accelerators built in the 1930s. Two of the main centers are Fermilab and CERN near Geneva, Switzerland. Large collaborations of scientists and engineers carry out research using state-of-the-art technologies in these laboratories.

BACK TO THE BUILDING BLOCKS

As scientists studied particle interactions in ever-increasing detail, they noticed slight differences in particle properties and behavior. Could this profusion of particles be explained by the existence of a small number of "building blocks" that joined together in many different combinations?

Chemists used a similar theory to explain interactions in chemical compounds. Fewer than one hundred chemical elements make up all the chemicals in the universe. Since the building blocks of atoms—protons, neutrons, and electrons—make up all the different atoms, scientists wondered whether "fundamental" building blocks of subatomic matter existed as well.

QUARKS AND OTHER PARTICLES

Today, most particle physicists agree that the universe is made up of two groups of particles: quarks and leptons. Other particles called gluons help the quarks stick together.

U.S. physicists Murray Gell-Mann and George Zweig suggested the theory of the quark in 1964. They both believed the quark was a building block of a proton or a neutron and was truly "elementary"—that is, it could not be divided. A few years later, the massive particle accelerator in Stanford, California,

Below: Electrons and neutrinos belong to the group of particles known today as leptons. This bubble-chamber photograph shows a neutrino (the white track on the right-hand side) interacting with an electron, then emerging as a neutrino again.

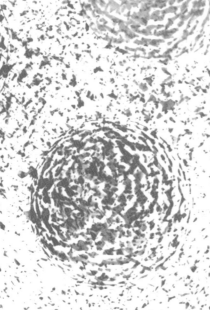

proved Gell-Mann and Zweig's theory. One surprising feature of this discovery was that quarks carried slight electrical charges, one-third or two-thirds that of a proton. This means that a proton is made up of three quarks. Electrons and neutrinos (the tiny particles created during nuclear reactions) are part of another group of "elementary" particles—leptons.

Today these ideas form part of what we call the Standard Model. In this model, everything is explained by using six different quarks that are matched by six leptons. The six quarks are named "up," "down," "strange," "charm," "bottom," and "top." For example, a proton is a combination of two "up" quarks and one "down" quark, while a neutron is a combination of one "up" and two "down" quarks.

In addition, four fundamental forces of nature affect all matter: gravity, electromagnetism, and the weak and strong interaction forces of atoms. All this very complicated physics proves a valuable point: Despite all the research scientists have done over the past two hundred years, and all the great leaps they have made in explaining the structure of the atom, there is still much to learn. Future research is bound to reveal even more secrets.

Key People

Murray Gell-Mann (b. 1929) was a U.S. physicist. Exceptionally talented at science even as a child, he enrolled in Yale University in New Haven, Connecticut, at age fifteen. At the time, many unidentified particles were being discovered—what we now call elementary particles. Gell-Mann focused on understanding and classifying these new particles. He grouped together particles that shared similar properties and left gaps in his tables for particles he believed existed but had not yet been discovered. (Dmitri Mendeleyev had done the same thing in the nineteenth century when he organized the known chemical elements.) Gell-Mann was awarded the Nobel Prize in physics in 1969 for his many valuable contributions to particle physics.

WHAT NEXT?

The path to splitting the atom was full of surprises, and the secrets learned so far were equally surprising. Identifying the "building-blocks" of matter proved fruitful, but research on these building blocks and what goes on inside subatomic particles continues. The building blocks of matter are atoms, and the building blocks of atoms are electrons, protons, and neutrons. The building blocks of protons and neutrons are quarks. And quarks? Perhaps that is for you to discover.

Below: A proton is made up of three quarks—two "up" quarks (blue) and one "down" quark (red).

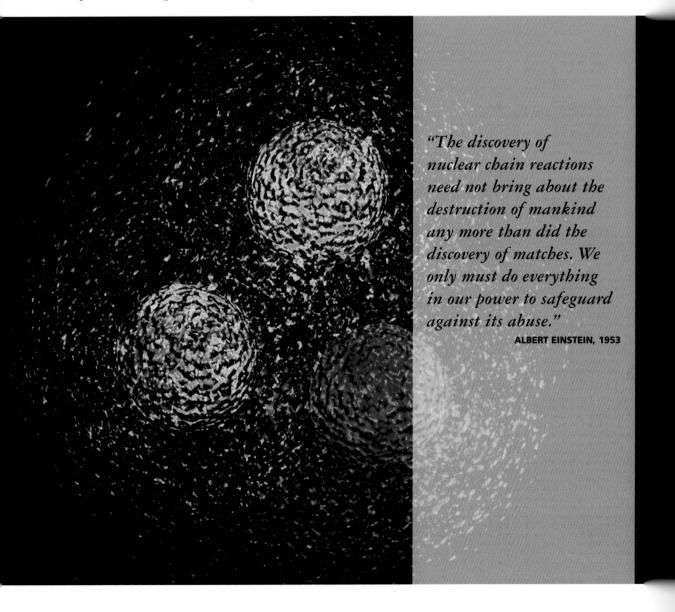

"*The discovery of nuclear chain reactions need not bring about the destruction of mankind any more than did the discovery of matches. We only must do everything in our power to safeguard against its abuse.*"

ALBERT EINSTEIN, 1953

TIME LINE

c. 400 B.C.	Democritus suggests that everything in the universe is made up of tiny, indivisible particles of matter
1789	Antoine Lavoisier draws up the first table of chemical elements
1803	John Dalton revives the atomic theory of matter
1815	William Prout suggests that atoms might be made up of even more fundamental "building blocks," which he calls protyles
1869	Dmitri Mendeleyev arranges the chemical elements into seven groups with similar properties, forming the basis of the periodic table
1895	Wilhelm Roentgen discovers X-rays
1896	Henri Becquerel discovers radioactivity in uranium
1897	J. J. Thomson discovers electrons
1898	Ernest Rutherford studies radiations from uranium and thorium and names them alpha and beta particles
1901	Marie Curie discovers polonium and radium
1905	Albert Einstein discovers the photoelectric effect
1911	Rutherford explains the results of the famous gold-foil experiment
1913	Henry Moseley's experiments with X-rays reveal information about atomic numbers and the positive charge of atoms
1919	Francis Aston discovers isotopes
1921	Rutherford suggests the name "proton" for the hydrogen nucleus
1922	Niels Bohr suggests a new "planetary" model of the atom
1925	Patrick Blackett photographs a collision between an alpha particle and a nitrogen nucleus
1927	G. P. Thomson demonstrates the wave nature of the electron
1929	The Cockcroft–Walton accelerator is built
1930	Ernest Lawrence builds the cyclotron; Paul Dirac proposes the existence of antimatter and antiparticles
1932	James Chadwick discovers the neutron
1934	Enrico Fermi discovers the neutrino
1939	Nuclear fission is achieved
1942	Fermi conducts the first controlled nuclear chain reaction
1945	The United States drops the first and only atomic bombs ever used in warfare on the Japanese cities of Hiroshima and Nagasaki
1950	American scientists begin work on the hydrogen bomb
1956	The world's first full-scale nuclear reactor is completed in England
1964	Murray Gell-Mann suggests the existence of "elementary" particles called quarks
2000	Evidence for a new elementary particle, the Higgs boson, is discovered at CERN

GLOSSARY

accelerator a machine that increases the speeds of beams of particles so they travel at speeds approaching the speed of light.

alpha particle a fast-moving helium nucleus, containing two protons and two neutrons, emitted during radioactive decay.

anode the positive end of an electrode.

antimatter an antiparticle; a subatomic particle with the same mass but opposite magnetic and electric properties of an ordinary subatomic particle. A union of the two causes instant annihilation of both.

antiparticle a subatomic particle not found in ordinary matter.

atom the smallest unit with the chemical properties of an element.

atomic bomb a nuclear weapon that uses fission to release energy in the form of an uncontrolled chain reaction.

atomic number the number of protons in the nucleus of an atom.

beta particle a high-speed particle, either an electron or a positron, emitted during radioactive decay.

bubble chamber a container of super-hot liquid in which an ionizing particle produces a string of bubbles that indicate its path.

cathode the negative end of an electrode.

cathode rays streams of electrons.

chain reaction a reaction that sustains itself by releasing energy or particles that cause more of the same kind of reaction.

cloud chamber a container of water vapor in which an ionizing particle creates a string of water droplets to indicate its path.

cosmic rays high-energy subatomic particles, usually protons and helium nuclei, traveling at speeds close to the speed of light.

cyclotron an accelerator that uses alternating electric fields to propel charged particles in a circular path within a constant magnetic field.

electrode an electrical conductor or end point through which a current enters or leaves.

electrolysis the use of an electric current to break down or separate substances into their component elements.

electromagnetic spectrum the complete range of differing wavelengths of mostly invisible radiation emitted by celestial bodies such as stars, pulsars, and supernovas.

electromagnetism one of the four fundamental forces of nature that governs ionic and subatomic particle interactions.

electron a massless subatomic particle and unit of negative electricity that orbits around the nucleus of an atom.

fission the splitting of an atom's nucleus.

fusion the combining of nuclei of light atoms (such as hydrogen) that forms a new element and releases enormous amounts of energy.

gamma rays a high-energy, dangerous form of electromagnetic radiation with very short wavelengths.

gluon the particle that carries the "strong interaction" force in an atom's nucleus. Gluons hold quarks together.

gravity the one-way attractive force between any two masses; one of the four fundamental forces of nature.

hydrogen bomb a fusion bomb that gets its power from the energy released when atomic nuclei of hydrogen combine.

infrared radiation a form of nonionizing electromagnetic radiation (that can nevertheless cause damage) with a longer wavelength than visible light; also called heat energy.

isotope an atom of an element that has the same properties as the basic element, but which contains a different number of neutrons and therefore has a different atomic mass.

kinetic energy the energy of motion.

lepton a fundamental subatomic particle—electrons and neutrinos are leptons.

mass a measure of how much matter something contains.

matter anything that has mass and takes up space.

molecule a group of atoms held together by chemical bonds.

neutrino a particle emitted during nuclear fission. Neutrinos have no electric charge and very low mass.

neutron a nuclear particle with mass but no electric charge.

nuclear reactor a large machine in which a controlled nuclear fission chain reaction is generated, often to produce electric power; also called a pile reactor.

nucleus the central core of an atom that contains the protons and neutrons. It is about ten thousand times smaller in diameter than the entire atom itself.

orbit the region, or shell, in which electrons move around the nucleus of an atom.

periodic table the table of chemical elements, arranged according to atomic number.

photoelectric effect the phenomenon of free electrons that are released when visible light strikes a metal surface.

photon the basic unit of light or other electromagnetic radiation. Atoms can emit or absorb energy in the form of photons

plasma a medium (with equal numbers of positively charged ions and electrons) that behaves somewhat like a gas but that conducts electricity and is affected by a magnetic field.

proton a positively charged particle found in the nucleus of atoms. The number of protons in a nucleus determines the chemical properties of an atom.

quantum mechanics the study of physics on tiny scales, such at subatomic levels.

quark an elementary particle that forms protons and neutrons.

radioactivity the spontaneous decay of one element into another radioactive element by the emission of invisible radiation.

relativity a collection of theories first suggested by Albert Einstein, that predicts and explains the relationships between light, time, mass, energy, and motion, and which forms the basis of our understanding of the universe.

spectrograph a instrument that records and maps the electromagnetic spectrum.

strong interaction one of the four fundamental forces of nature; the subatomic force that binds together nuclear quarks.

thermal energy heat energy.

tokamak a Russian-designed, doughnut-shaped fusion reactor that uses powerful, precisely aligned magnetic fields to keep heated plasma from touching any of its surfaces.

ultraviolet radiation a form of ionizing electromagnetic radiation with a shorter, penetrating wavelength that damages living tissues by causing molecular changes.

vacuum a region that contains no free matter.

volt a unit of electrical energy.

weak interaction one of the four fundamental forces of nature; responsible for the behavior of neutrinos, leptons, and particle decay such as beta radiation.

X-rays a form of electromagnetic radiation with very short wavelengths and high energy.

FURTHER INFORMATION

BOOKS

Bodanis, David. *E=mc²: A Biography of the World's Most Famous Equation*. Berkley (2001).

Bradley, David. *Atoms and Elements*. The Young Oxford Library of Science (series). Oxford University Press (2003).

Heinrichs, Ann. *Albert Einstein. Trailblazers of the Modern World* (series). World Almanac Library (2002).

Heilbron, John L. *Ernest Rutherford and the Explosion of Atoms*. Oxford University Press (2003).

Lawton, Clive A. *Hiroshima: The Story of the First Atom Bomb*. Candlewick Press (2004).

Nardo, Don. *Atoms*. The Kidhaven Science Library, Kidhaven Press (2001).

Oxlade, Chris. *Atoms in Action*. Heinemann (2002).

Parker, Steve. *Nuclear Energy. Science Files: Energy* (series). Gareth Stevens (2004).

WEB SITES

http://nobelprize.org/
Visit the official Nobel Prize home page. Follow the links to play the Peace Doves game.

http://public.web.cern.ch/Public/Welcome.html
Check out the home page for CERN, the world's largest particle physics laboratory.

www.fnal.gov/pub/about/tour/index.html
Take a virtual tour of Fermilab.

www.howstuffworks.com/atom.htm
Learn how atom smashers work and much more.

www.lbl.gov/abc/
Search for diagrams, definitions, and information about everything from the basic structure of the atom to radioactivity and the cosmic connection.

www.nrc.gov/reactors/operating/map-power-reactors.html
View a United States map showing nuclear power plant locations.

www.pgs.ca/www.bombsaway.ca/bunker/factsheet.html
Get the facts on the destructive power of nuclear weapons.

http://science.csustan.edu/perona/3100/3070m04.htm#sample
Test your knowledge with an instant-feedback quiz on nuclear properties.

INDEX